Practice Makes Perfect

Measurement

GRADES 1&2

Authors

Teacher Created Materials Staff
and Mary Rosenberg

Managing Editor
Karen J. Goldfluss, M.S. Ed.

Editor-in-Chief
Sharon Coan, M.S. Ed.

Cover Artist
Barb Lorseyedi

Illustrator
Howard Chaney

Art Coordinator
Kevin Barnes

Art Director
CJae Froshay

Imaging
Rosa C. See

Product Manager
Phil Garcia

Publishers
Rachelle Cracchiolo, M.S. Ed.
Mary Dupuy Smith, M.S. Ed.

Teacher Created Materials, Inc.
6421 Industry Way
Westminster, CA 92683
www.teachercreated.com
ISBN-0-7439-3319-2

Reprinted, 2003
Made in U.S.A.

Table of Contents

Introduction

The old adage "practice makes perfect" can really hold true for your child and his or her education. The more practice and exposure your child has with concepts being taught in school, the more success he or she is likely to find. For many parents, knowing how to help your children can be frustrating because the resources may not be readily available. As a parent it is also difficult to know where to focus your efforts so that the extra practice your child receives at home supports what he or she is learning in school.

This book has been designed to help parents and teachers reinforce basic skills with children. *Practice Makes Perfect* reviews basic math skills for children in grades 1 and 2. The math focus is on linear and liquid measurement. While it would be impossible to include all concepts taught in grades 1 and 2 in this book, the following basic objectives are reinforced through practice exercises. These objectives support math standards established on a district, state, or national level. (Refer to the Table of Contents for the specific objectives of each practice page.)

- comparing sizes of objects
- estimating, measuring, and comparing lengths using nonstandard units
- measuring distances from one point to another using nonstandard units
- estimating and measuring lengths in inches and centimeters
- measuring and drawing lines
- naming objects that can be measured in inches, centimeters, meters, or yards
- using centimeter rulers and inch rulers
- using a yard stick and a meter stick
- measuring and comparing liquids (cups, pints, quarts, gallons, liters)

There are 31 practice pages organized sequentially, so children can build their knowledge from more basic skills to higher-level math skills. (**Note:** Have children show all work where computation is necessary to solve a problem. For multiple choice responses on practice pages, children can fill in the letter choice or circle the answer.) Following the practice pages are four test practices. These provide children with multiple-choice test items to help prepare them for standardized tests administered in schools. As your child completes each test, he or she can fill in the correct bubbles on the optional answer sheet provided on page 47. To correct the test pages and the practice pages in this book, use the answer key provided on page 48.

How to Make the Most of This Book

Here are some useful ideas for optimizing the practice pages in this book:

- Set aside a specific place in your home to work on the practice pages. Keep it neat and tidy with materials on hand.
- Set up a certain time of day to work on the practice pages. This will establish consistency. Look for times in your day or week that are less hectic and more conducive to practicing skills.
- Keep all practice sessions with your child positive and more constructive. If the mood becomes tense, or you and your child are frustrated, set the book aside and look for another time to practice with your child.
- Help with instructions if necessary. If your child is having difficulty understanding what to do or how to get started, work through the first problem with him or her.
- Review the work your child has done. This serves as reinforcement and provides further practice.
- Allow your child to use whatever writing instruments he or she prefers. For example, colored pencils can add variety and pleasure to drill work.
- Pay attention to the areas in which your child has the most difficulty. Provide extra guidance and exercises in those areas. Allowing children to use drawings and manipulatives, such as coins, tiles, game markers, or flash cards, can help them grasp difficult concepts more easily.
- Look for ways to make real-life applications to the skills being reinforced.

Practice 1

1. Which clock is the tallest?

(A) 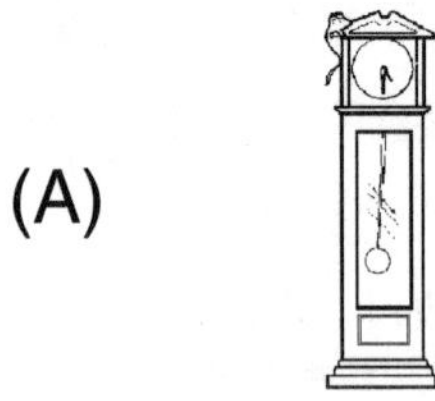 (B) 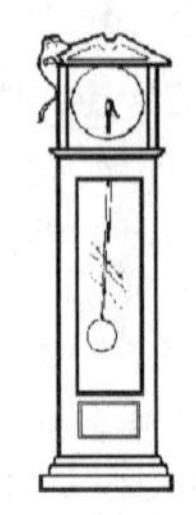 (C) 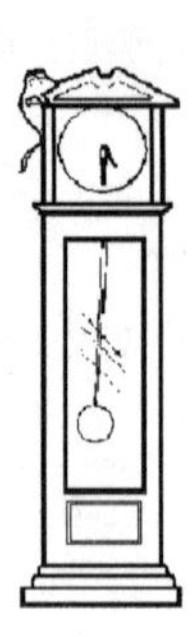

2. Which spoon is the shortest?

(A) 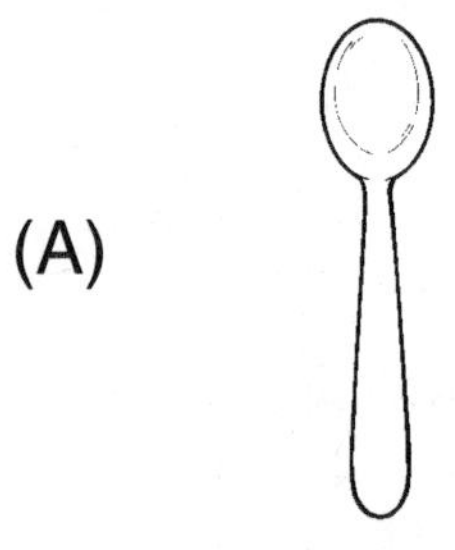 (B)  (C) 

3. Which cup is the tallest?

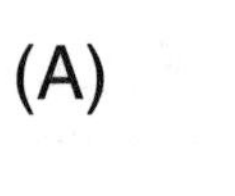

(A)  (B)  (C) 

4. Look at the forks.

(1)   (2) 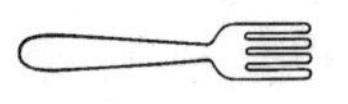 (3) 

Which group shows the forks placed from ***shortest*** **to** ***longest***?

(A) 2, 1, 3 (B) 3, 1, 2

(B) 2, 3, 1 (D) 1, 2, 3

5. Which tent is the smallest?

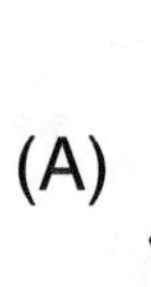

(A)  (B)  (C) 

Practice 2

1. Which group lists the items from ***shortest*** **to** ***longest***?

(A) a chalkboard, a pencil, your shoe, a chair

(B) a pencil, your shoe, a chair, a chalkboard

---

2. Which group lists the items from ***shortest*** **to** ***longest***?

(A) a paper clip, a bicycle, a train, a ruler

(B) a paper clip, a ruler, a bicycle, a train

---

3. List the items from ***shortest*** **to** ***tallest*** (as they would look in real life).

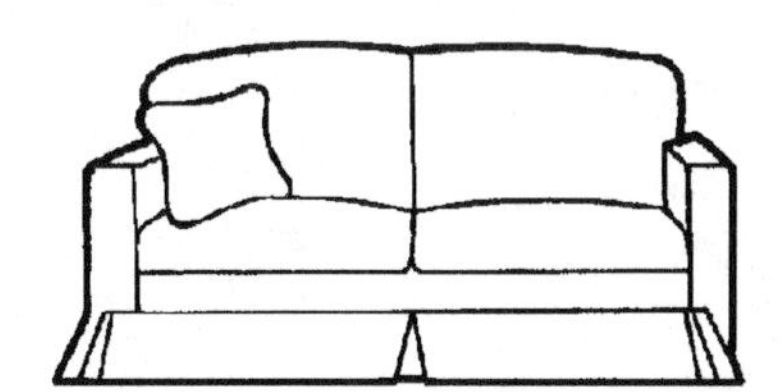

____________ ____________ ____________

---

4. List the items from ***tallest*** **to** ***shortest*** (as they would look in real life).

____________ ____________ ____________

Practice 3

1. About how many paper clips long is the sandal?

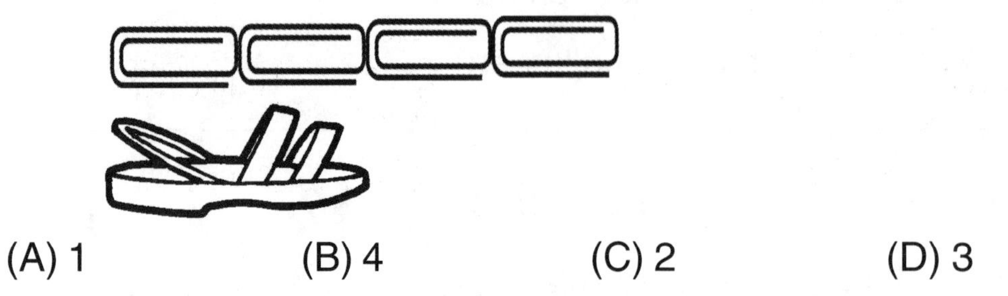

(A) 1 (B) 4 (C) 2 (D) 3

2. About how many paper clips long is the leaf?

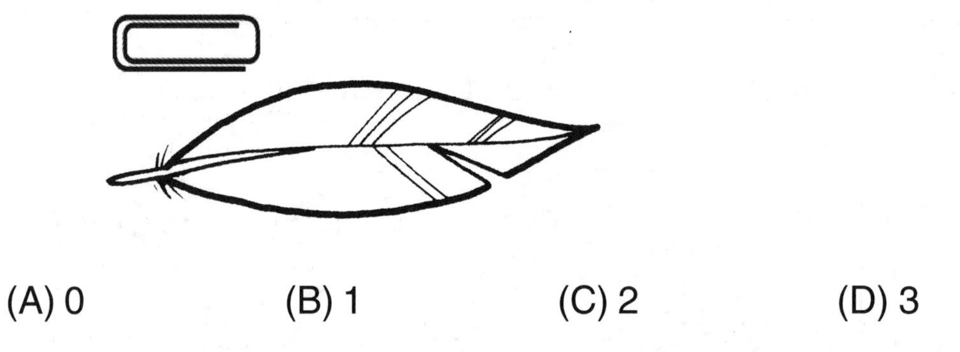

(A) 0 (B) 1 (C) 2 (D) 3

3. About how many paper clips long is the pencil?

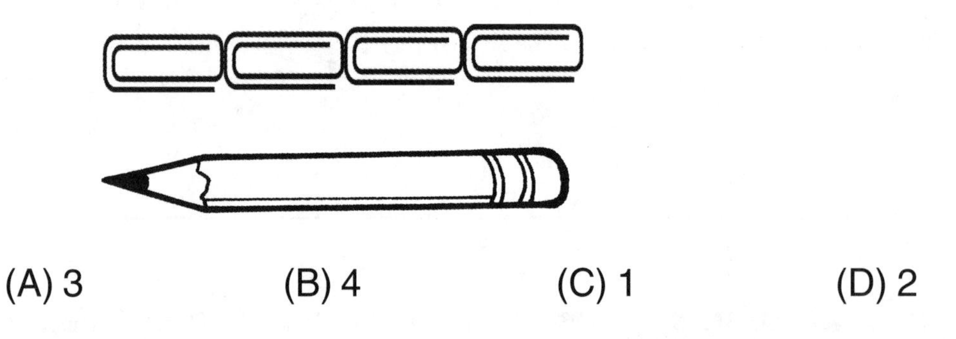

(A) 3 (B) 4 (C) 1 (D) 2

4. About how many paper clips long is the comb?

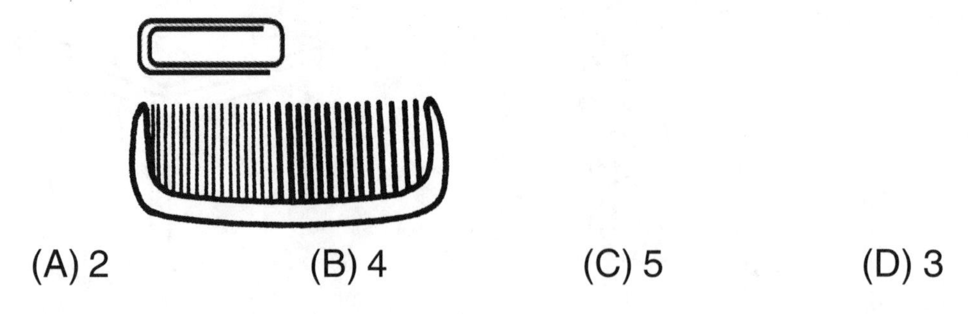

(A) 2 (B) 4 (C) 5 (D) 3

Practice 4

About how many paper clips long is each item below?

1.

It is ______ long.

2.

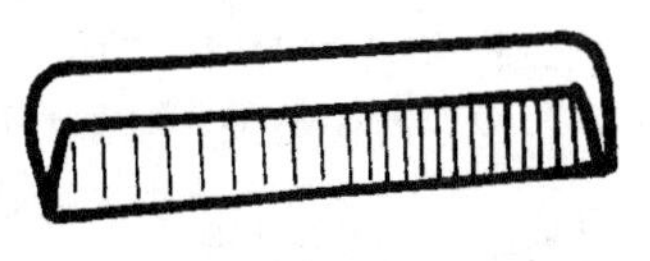

It is ______ long.

3.

It is ______ long.

4.

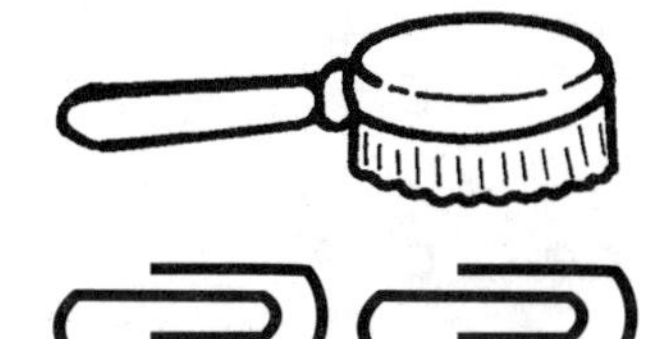

It is ______ long.

5. About how many crabs can fit side by side along the line? ________

6. About how many bowls can fit side by side along the line? ________

Practice 5

A hand can be used to measure things. Holding the hand open and fingers together is called "palm measuring." Look at the picture and hold your hand this way. Turn your palm down and place your hand in the box. Have someone trace around your hand. Now use palm measuring to measure these items.

palm

Directions: Place a hand at one edge of the item. Then place the other hand next to the first hand and "leap frog" the first hand over the second hand. Continue to "leap frog" until you reach the end of the item you are measuring.

1. chair (top to bottom)

about ______ palms

2. door (side to side)

about ______ palms

3. bookcase (top to bottom)

about ______ palms

4. kitchen counter (top to bottom)

about ______ palms

5. window (side to side)

about ______ palms

6. shoe (heel to toe)

about ______ palms

Practice 6

1. Circle the shortest line.

A. ____________________

B. __________

C. _______________

2. Circle the shortest line.

A. __________

B. _______________

C. ____________________

3. Circle the shortest line.

A. _______________

B. ____________________

C. __________

4. Use one of the items on pages 43 or 45 to measure the length of your paper.

I used ________________________________.

My paper is ___________________ long.

5. Use one of the items on pages 43 or 45 to measure the length of your pencil.

I used ______________________________.

My pencil is __________________ long.

6. Circle the longest line.

A. ____________________

B. __________

C. _______________

7. Circle the longest line.

A. __________

B. _______________

C. ____________________

8. Circle the longest line.

A. _______________

B. ____________________

C. ____________

9. Use one of the items on pages 43 or 45 to measure the length of an eraser.

I used ________________________________.

My eraser is ___________________ long.

10. Use one of the items on pages 43 or 45 to measure the length of a crayon.

I used ______________________________.

My crayon is _________________ long.

Practice 7

Use the paper clips on page 45 to measure about how long each line is. Line up the paper clips as shown in the sample. In the box, write the measurement to the nearest paper clip.

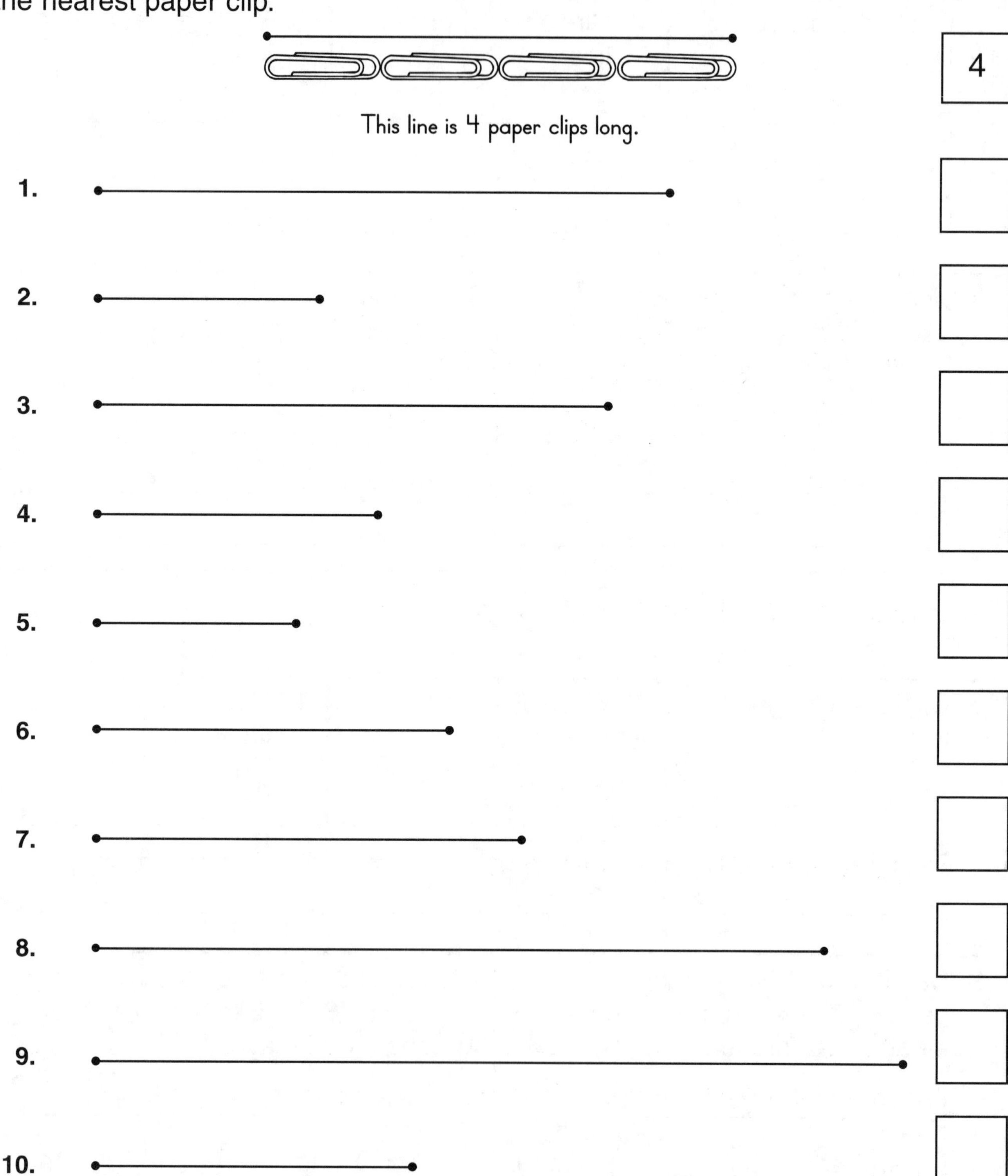

Practice 8

1. About how many units is it from the start of Path 1 to the end of Path 2?

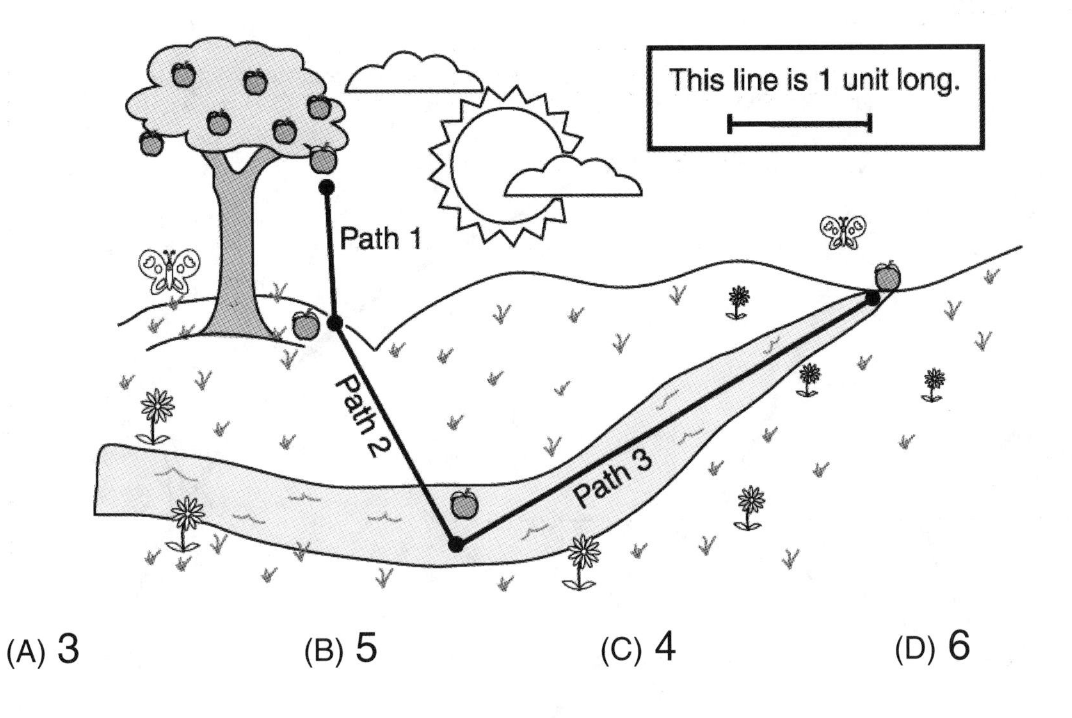

(A) 3 (B) 5 (C) 4 (D) 6

---

2. Which path is about 2 units long?

____

---

3. Which path is about 3 units long?

____

---

4. About how many units is it from the start of Path 2 to the end of Path 3?

____

Practice 9

The inchworms in this picture are about one inch long. Use the picture to answer the questions below.

Each inchworm has a leaf to measure.

1. Which inchworm has the longest leaf to measure? ______________________

2. Which inchworm has the shortest leaf to measure? ______________________

3. Which two worms are measuring the same-sized leaf?

   ______________________ and ______________________

4. Which leaf is longer, Kim's or Lori's? ______________________

5. If they both move at the same speed, will Juan or Kim finish measuring the leaf first? ______________________

Practice 10

How much is an inch? Look at the worms on the leaf. In the box, write the letter of the worm that you think is an inch long.

1.

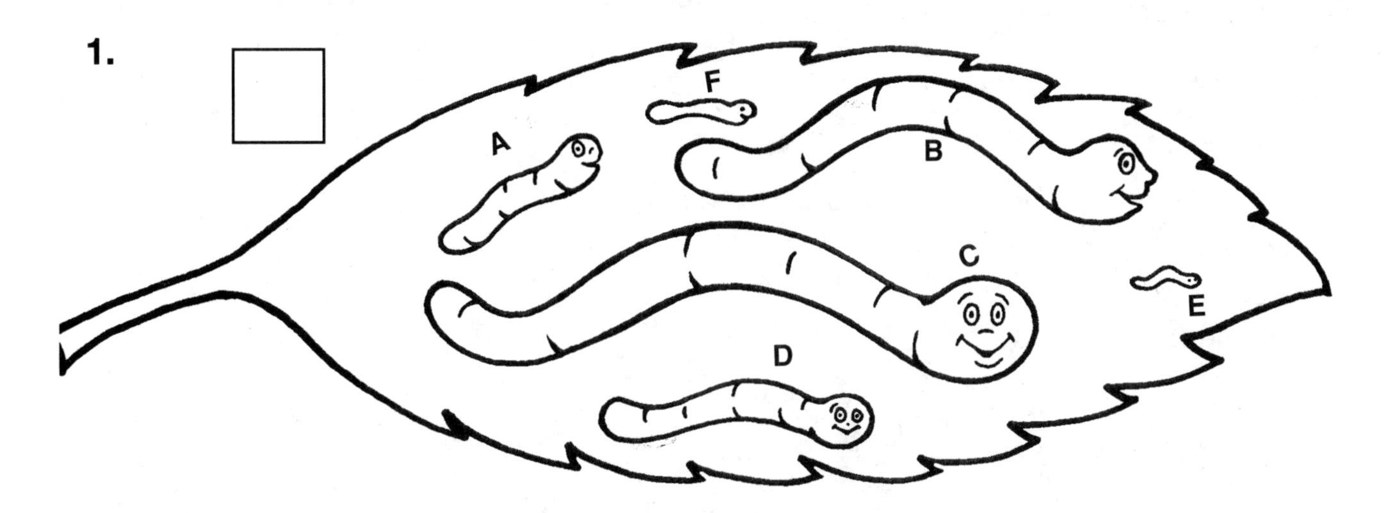

Now, cut out and use the inchworms on page 43 to measure the objects below. Write each length in the box.

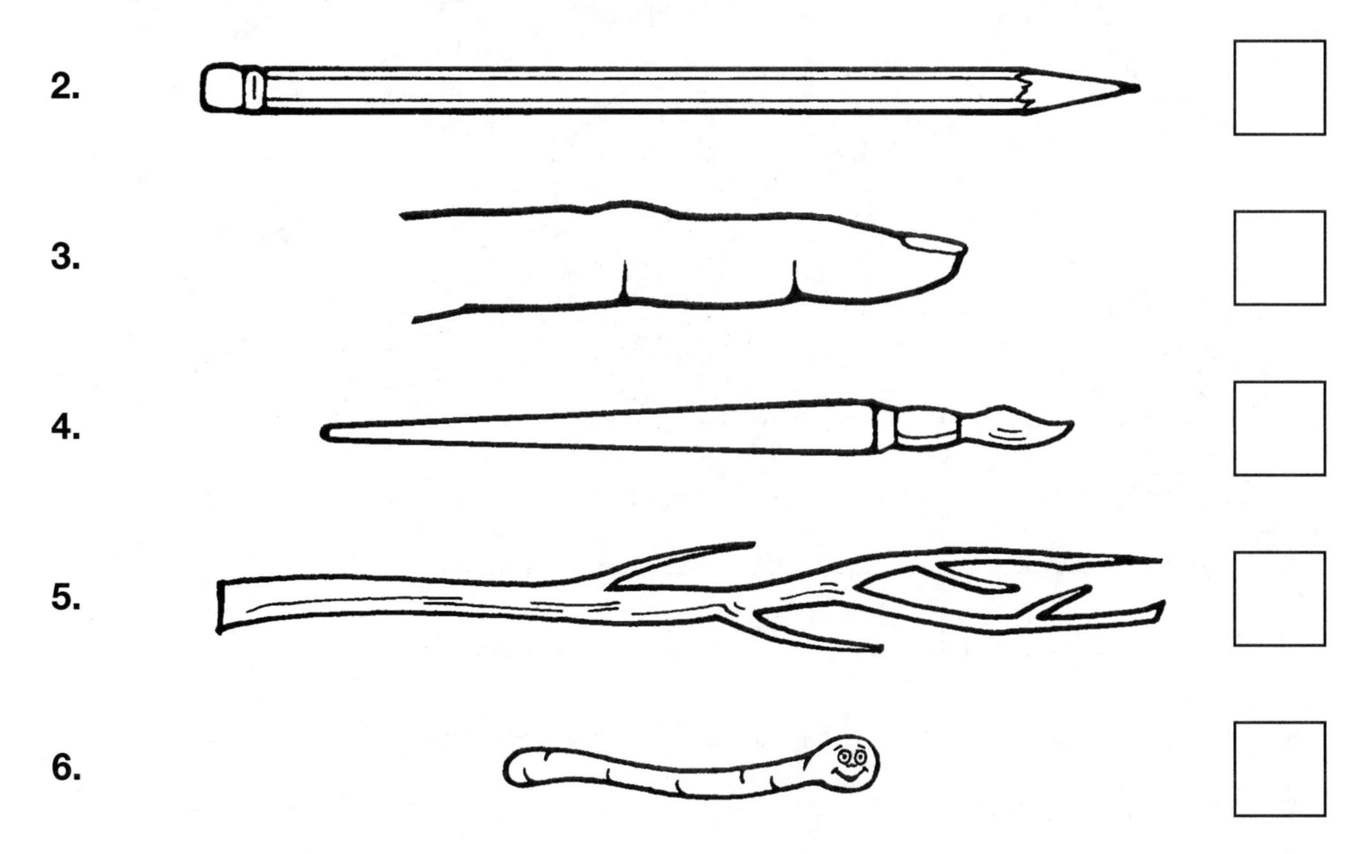

Practice 11

Decide if the items in the bubbles can or cannot be measured in inches. If you think using inches to measure the item is best, color the bubble and matching "YES" box yellow. If you think using inches is not a good choice, color the bubble and matching "NO" box green. Then, answer the questions at the bottom of the page.

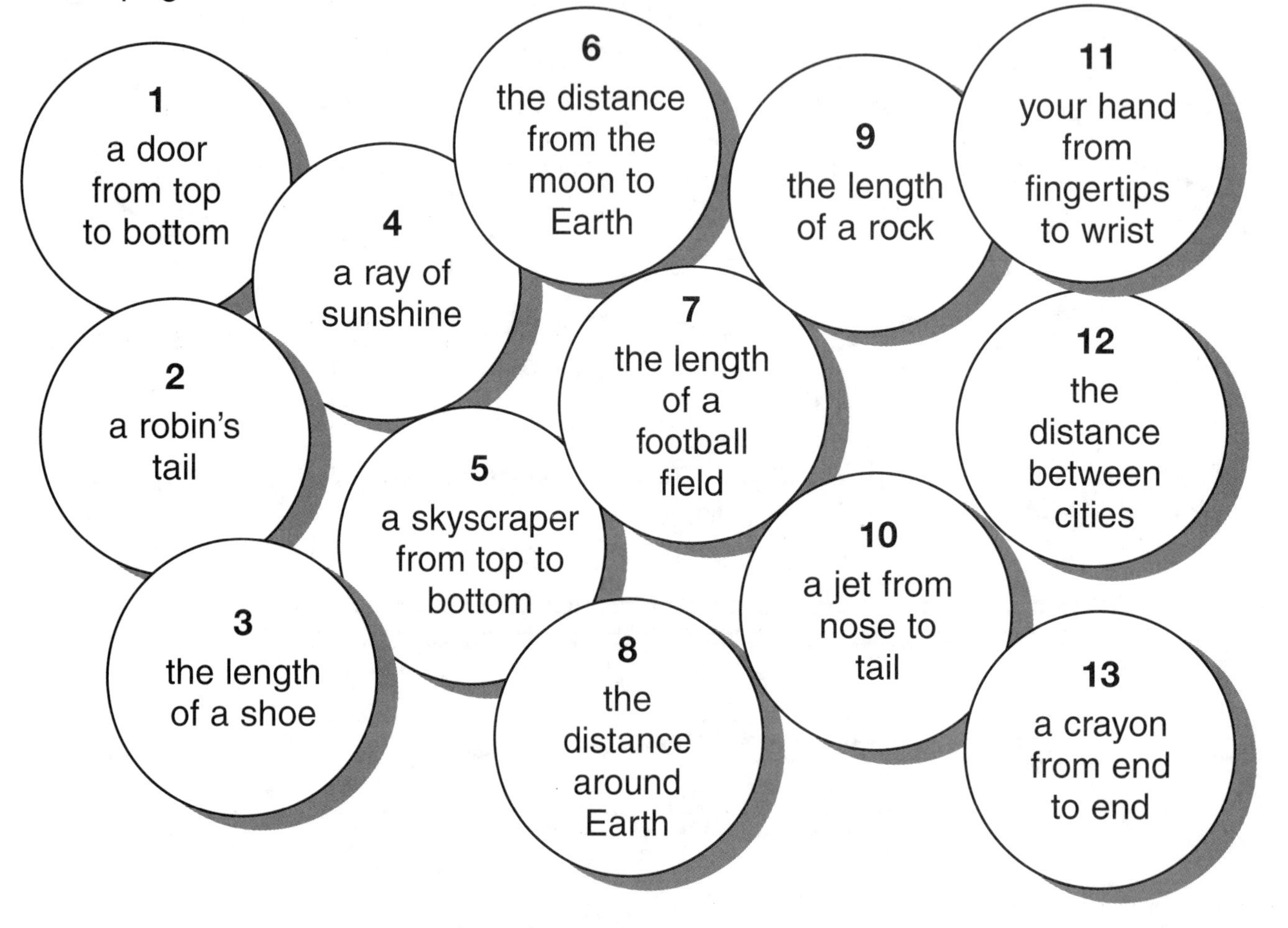

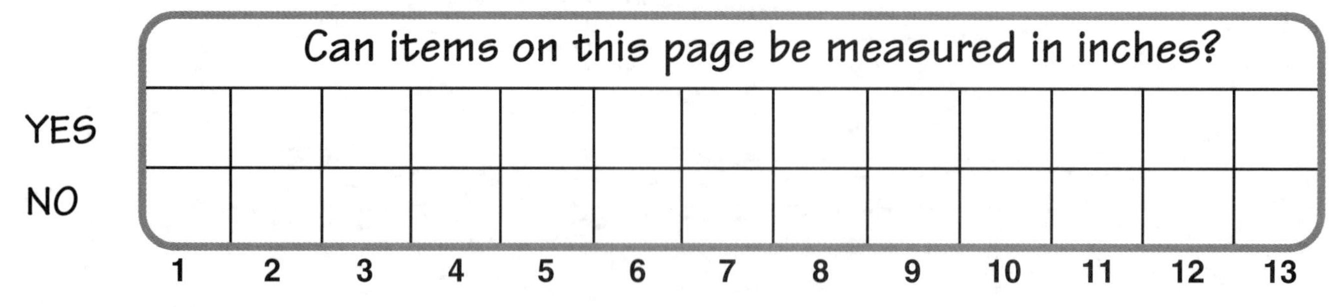

Can items on this page be measured in inches?

	1	2	3	4	5	6	7	8	9	10	11	12	13
YES													
NO													

1. How many items on this page can be measured in inches? ____________
2. How many items on this page cannot be measured in inches? ____________
3. Are there more items on this page that can be measured or that cannot be measured in inches? ________________ How many more? ____________

Practice 12

This paper can be measured in inches. The inchworms on page 43 are each one inch long. Cut them out and lay them end to end to find the following measurement of this paper:

1. top to bottom about _________ inches long

2. side to side about _________ inches long

3. one corner to the opposite corner about _________ inches long

Now find 10 items that can be measured using the inchworms. Then, on the lines, write the items you chose and the number of inchworms you used to measure each item.

Item	Number of Inchworms I Used
________________________	________
________________________	________
________________________	________
________________________	________
________________________	________
________________________	________
________________________	________
________________________	________
________________________	________
________________________	________

Practice 13

1. About how many paper clips long is the log?

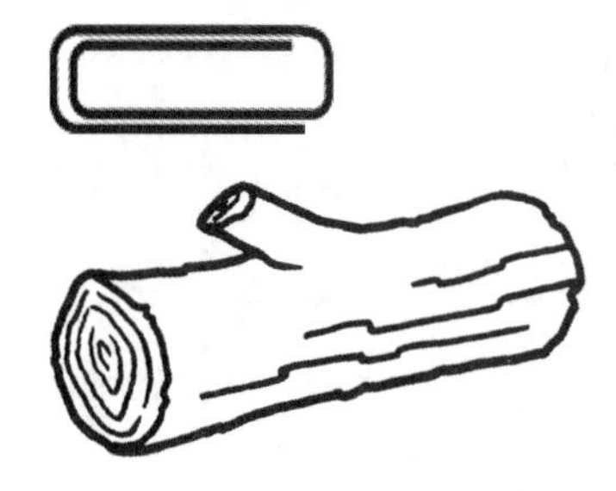

____

2. Measure the item. About how long is it?

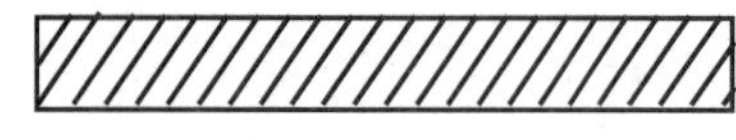

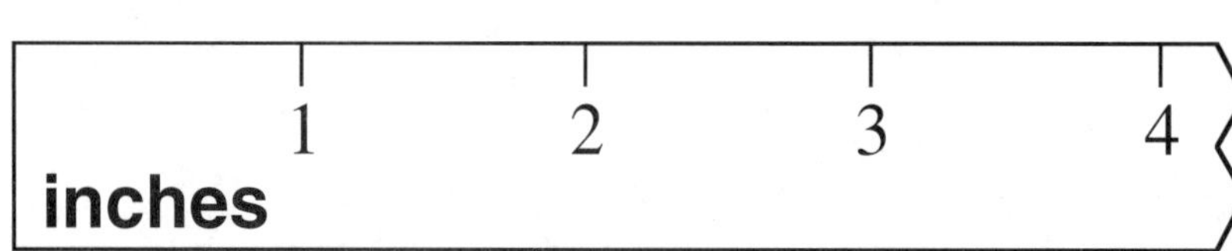

(A) about 2 inches
(B) about 3 inches
(C) about 1 inch
(D) about 4 inches

3. About how many paper clips long is the rake?

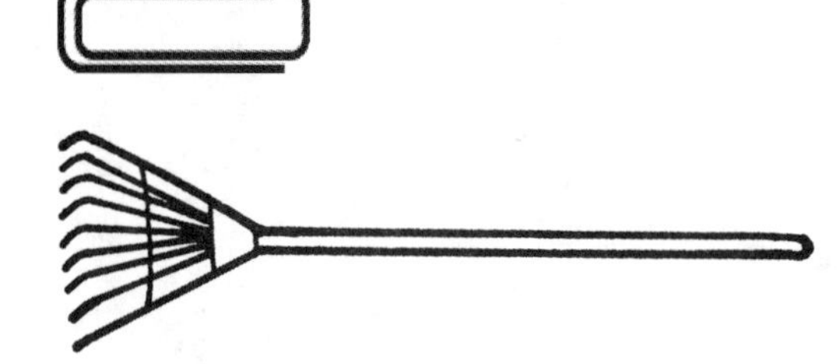

____

4. Measure the item. About how long is it?

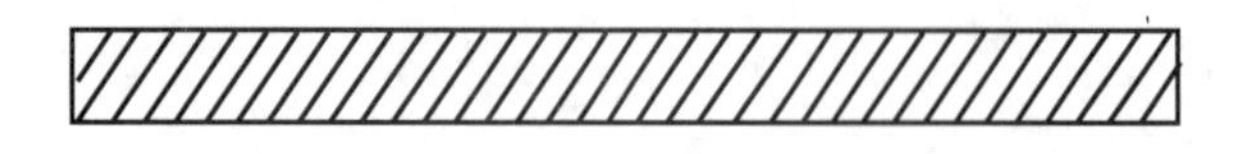

____

1 2 3 4
inches

Practice 14

The inchworms on page 43 are each one inch long. Use the inchworms to measure about how long each bar is from end to end. Write your answers on the lines.

1.

I used ______________ inchworms to measure. The bar is about ______________ inches long.

2.

I used ______________ inchworms to measure. The bar is about ______________ inches long.

3.

I used ______________ inchworms to measure. The bar is about ______________ inches long.

4.

I used ______________ inchworms to measure. The bar is about ______________ inches long.

5.

I used ______________ inchworms to measure. The bar is about ______________ inches long.

6.

I used ______________ inchworms to measure. The bar is about ______________ inches long.

Practice 15

Follow the directions on page 39 to make a ruler. Use the ruler to measure each inchworm. Write how many inches each worm is on the line.

1. ________ inches

2. ________ inch

3. ________ inches

4. ________ inches

5. ________ inches

6. ________ inches

7. ________ inches

8. ________ inches

Practice 16

Follow the directions on page 39 to make a ruler. Use the ruler to measure and draw lines next to the lengths listed below. Begin your line at the star.

1. 7 inches *

2. 4 inches *

3. 1 inch *

4. 3 inches *

5. 6 inches *

6. 2 inches *

7. 5 inches *

Practice 17

When measuring with a ruler, place the beginning of the ruler directly underneath the beginning of the object being measured. Measure to the end of the object and record the number.

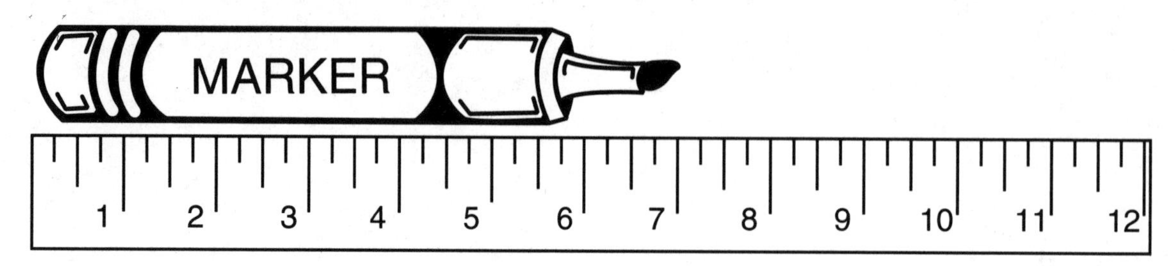

The marker is 7 inches long.

Directions: Measure each item to the nearest inch.

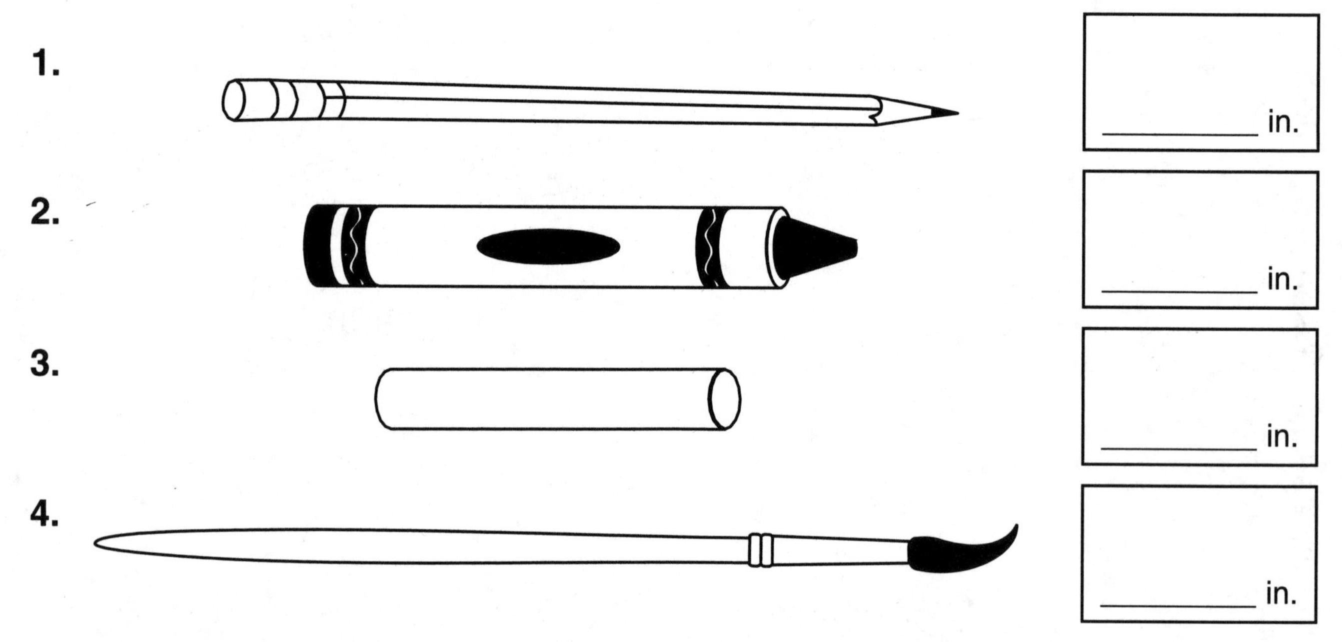

Directions: Use a ruler to measure the following actual (or real) body parts. Circle whether it is less than a foot (less than 12") or more than a foot (more than 12").

5.	less	more	8.	less	more
6.	less	more	9.	less	more
7.	less	more	10.	less	more

Practice 18

Centimeters are used to measure items in the metric measuring system. One centimeter is less than half an inch. Look at the centimeter bar below each item. Each part of the bar is one centimeter. Use the bars to find how many centimeters each picture is from end to end.

1.

________ centimeters

2.

________ centimeters

3.

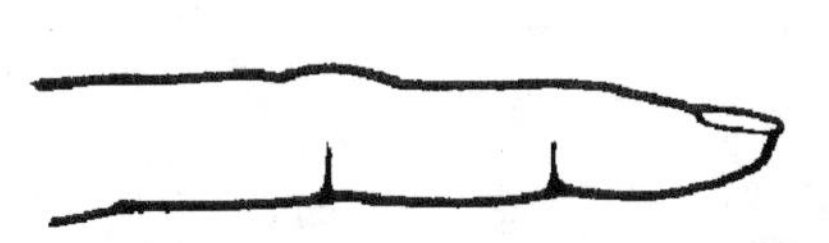

________ centimeters

4.

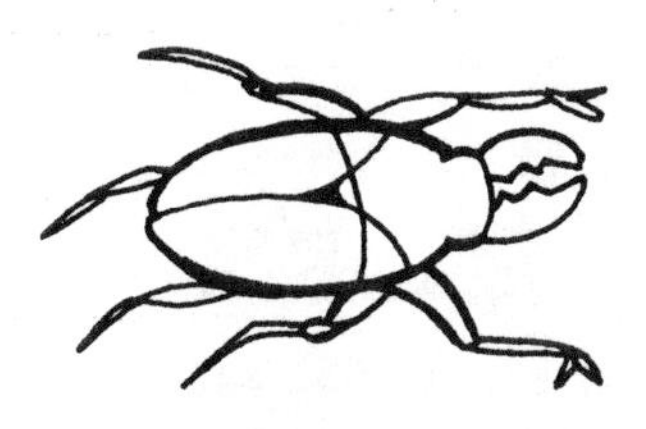

________ centimeters

5.

________ centimeters

6.

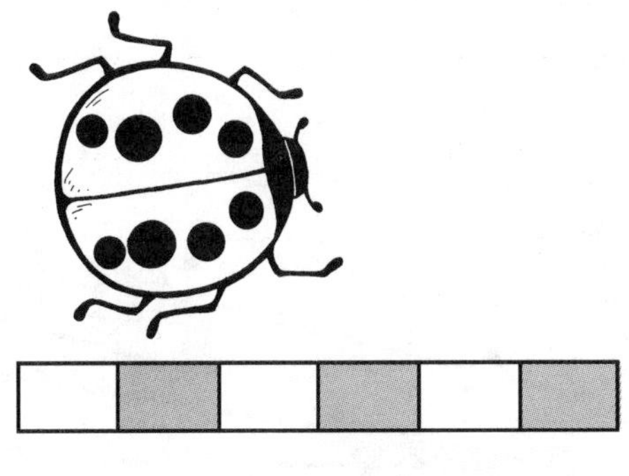

________ centimeters

Practice 19

To measure in centimeters, place the beginning of a centimeter ruler underneath the beginning of the item being measured. Measure to the end of the item. The short way to write centimeters is "cm."

Use the first centimeter strip on page 41 to measure each item below to the nearest centimeter (cm). Write the answer on the line.

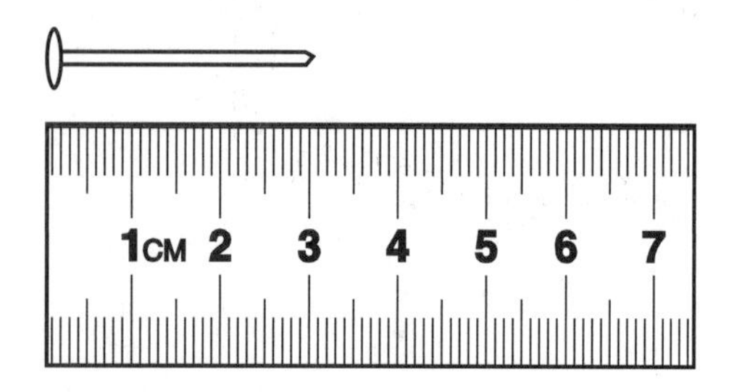

The nail is 3 centimeters or 3 cm long.

Directions: Measure each item below to the nearest centimeter. Write the answer on the line.

1. ______ cm

2. ______ cm

3. ______ cm

4. ______ cm

5. ______ cm

6. ______ cm

7. ______ cm

8. ______ cm

Practice 20

Use the centimeter ruler on page 41 to measure each of the frog's plants to the nearest centimeter. Write your answer below each plant.

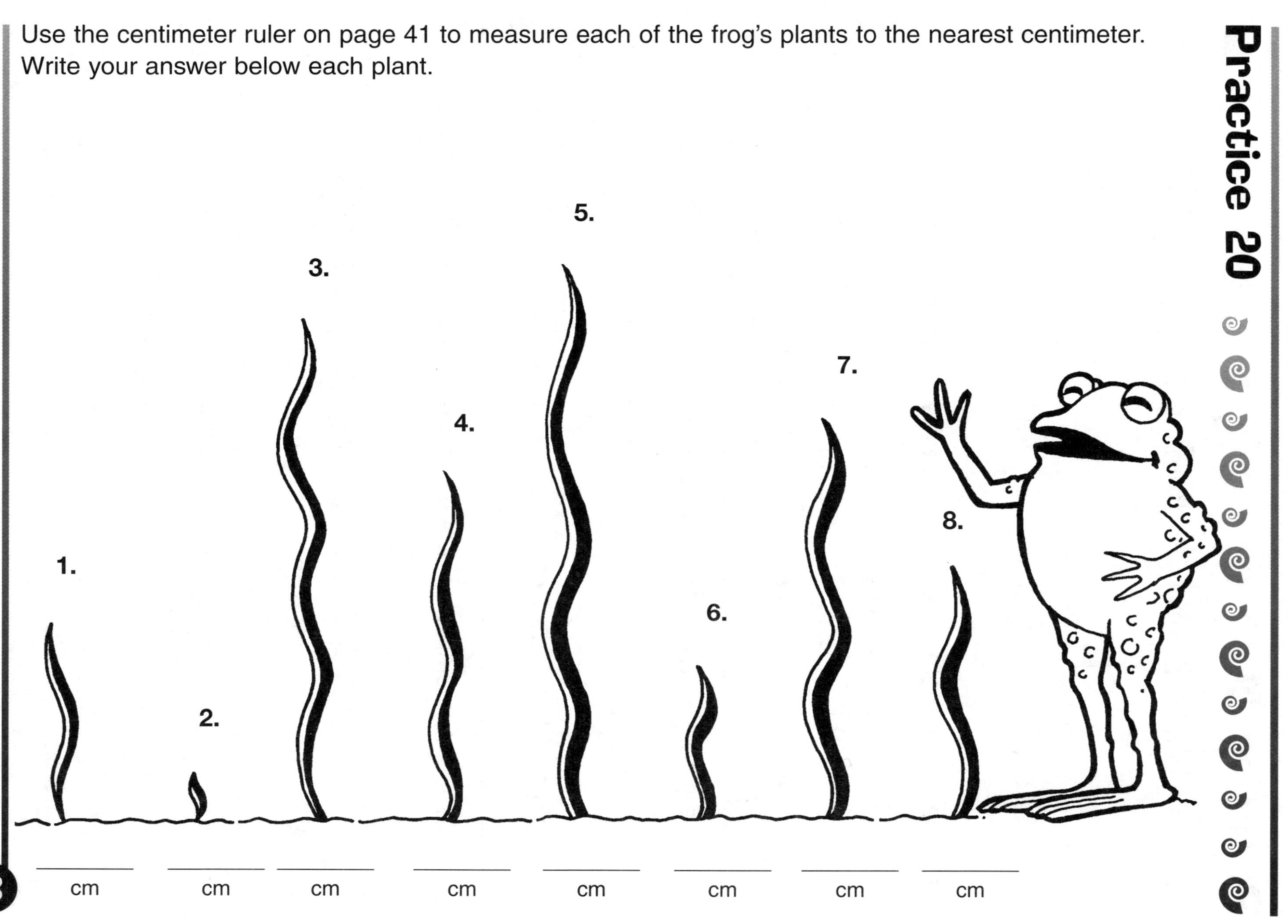

Practice 21

1. Use a centimeter ruler. Measure the length of the box below.

(A) 2 cm (B) 3 cm (C) 4 cm (D) 1 cm

---

2. Use a centimeter ruler. Measure the length of the box below.

(A) 4 cm (B) 6 cm (C) 7 cm (D) 5 cm

---

3. Use a centimeter ruler. Measure the length of the box below.

(A) 8 cm (B) 7 cm (C) 9 cm (D) 10 cm

---

4. Use a centimeter ruler. Measure the length of the box below.

(A) 8 cm (B) 6 cm (C) 7 cm (D) 9 cm

---

5. Use a centimeter ruler. Measure the length of the box below.

(A) 13 cm (B) 10 cm (C) 11 cm (D) 12 cm

Practice 22

Measure It, Draw It

Use the first centimeter ruler on page 41 to measure and draw lines next to the lengths listed below. Begin your line at the star.

1. 7 centimeters *

2. 4 centimeters *

3. 12 centimeters *

4. 8 centimeters *

5. 3 centimeters *

6. 9 centimeters *

7. 1 centimeter *

8. 15 centimeters *

Practice 23

Estimate how far each ladybug flew. Use centimeters for your estimate. Record your estimations. Measure the actual length of each ladybug's path using a centimeter ruler. Record the measurement. Were the estimates too long, too short, just right?

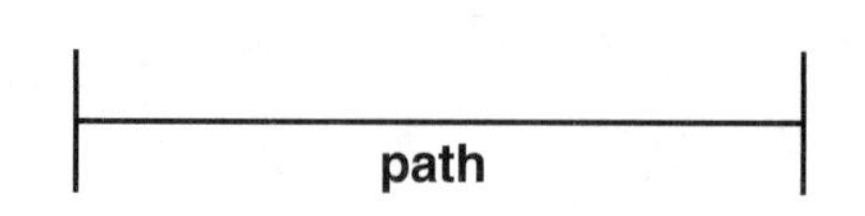

1. Estimate ______________ Actual ________________

path

2. Estimate ______________ Actual ________________

path

3. Estimate ______________ Actual ________________

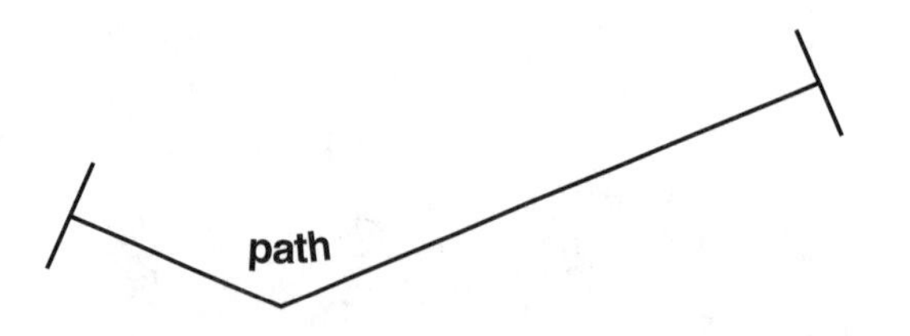

4. Estimate ______________ Actual ________________

5. Which ladybug went the farthest? ____________________________

Practice 24

Meters are used to measure lengths of items that are too large to be measured in centimeters. How long is a meter? Make your own meter ruler by cutting out and attaching the strips on page 41. Then, use the meter ruler to help you decide whether the actual (real) items shown in the pictures are shorter or taller than one meter in real life. On the lines, write **shorter** or **taller**.

1.  ____________	2.  ____________	3.  ____________
4. 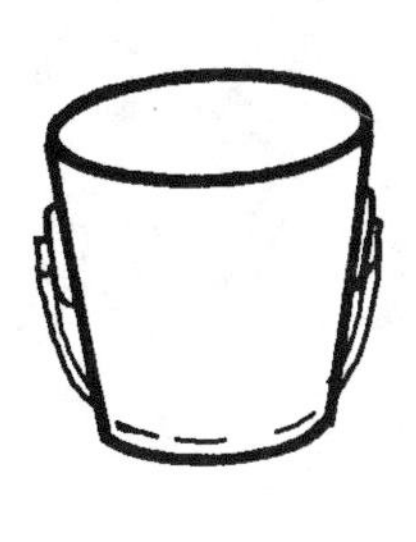 ____________	5.  ____________	6.  ____________
7.  ____________	8.  ____________	9. 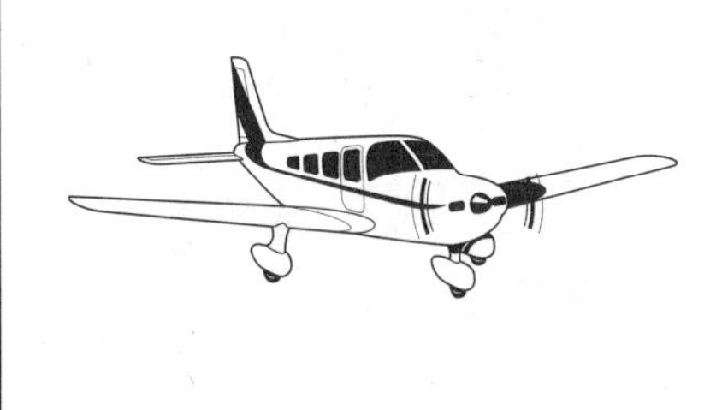 ____________

Practice 25           

It takes 100 centimeters to equal one meter. Decide if the actual (real) items pictured below should be measured in centimeters or meters. On the line write "cm" for centimeters or "m" for meters.

1.

____

2.

____

3.

____

4.

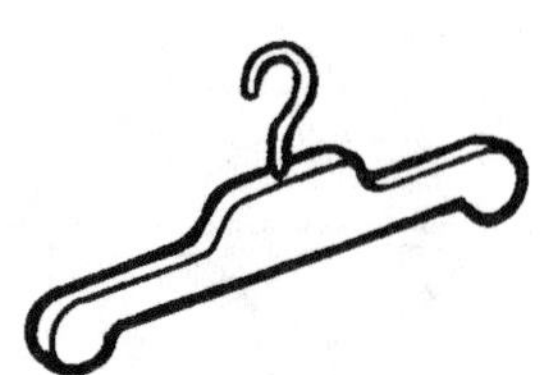

____

5.

____

6.

____

7.

____

8.

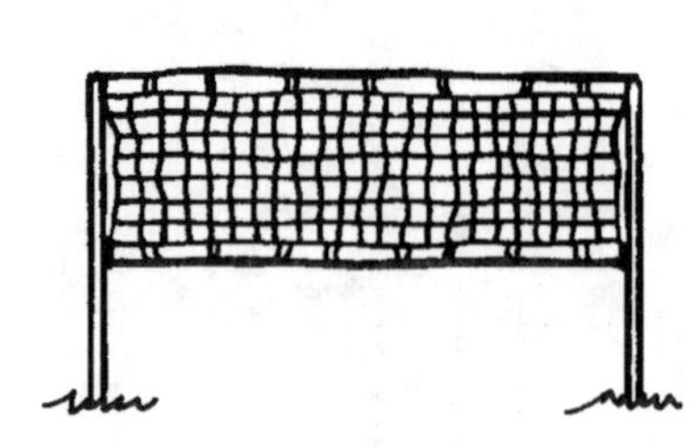

____

9.

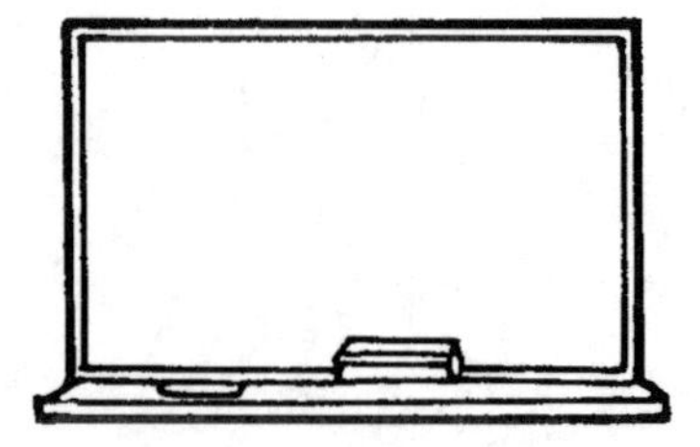

____

Practice 26

If you place three 12-inch rulers end to end, the measurement would be 36 inches, or one yard. A yard stick is used to measure items that are too long to measure in inches or feet. Make a yard stick using the patterns on page 39. Then, use it to measure ten large items in your house or at school. On the chart below, write the item you measured and about how many yards long it is.

Item	About How Many Yards Long
______________________	__________
______________________	__________
______________________	__________
______________________	__________
______________________	__________
______________________	__________
______________________	__________
______________________	__________
______________________	__________
______________________	__________

Practice 27

Use the numbers 1–4 to place the items in order from the items that would hold the least amount to the items that would hold the greatest amount of liquid.

1.

quart ______ cup ______ gallon ______ pint ______

What should be used (A, B, C, or D) to fill the item shown at the top of each box? Fill in the circle under the appropriate measurement container.

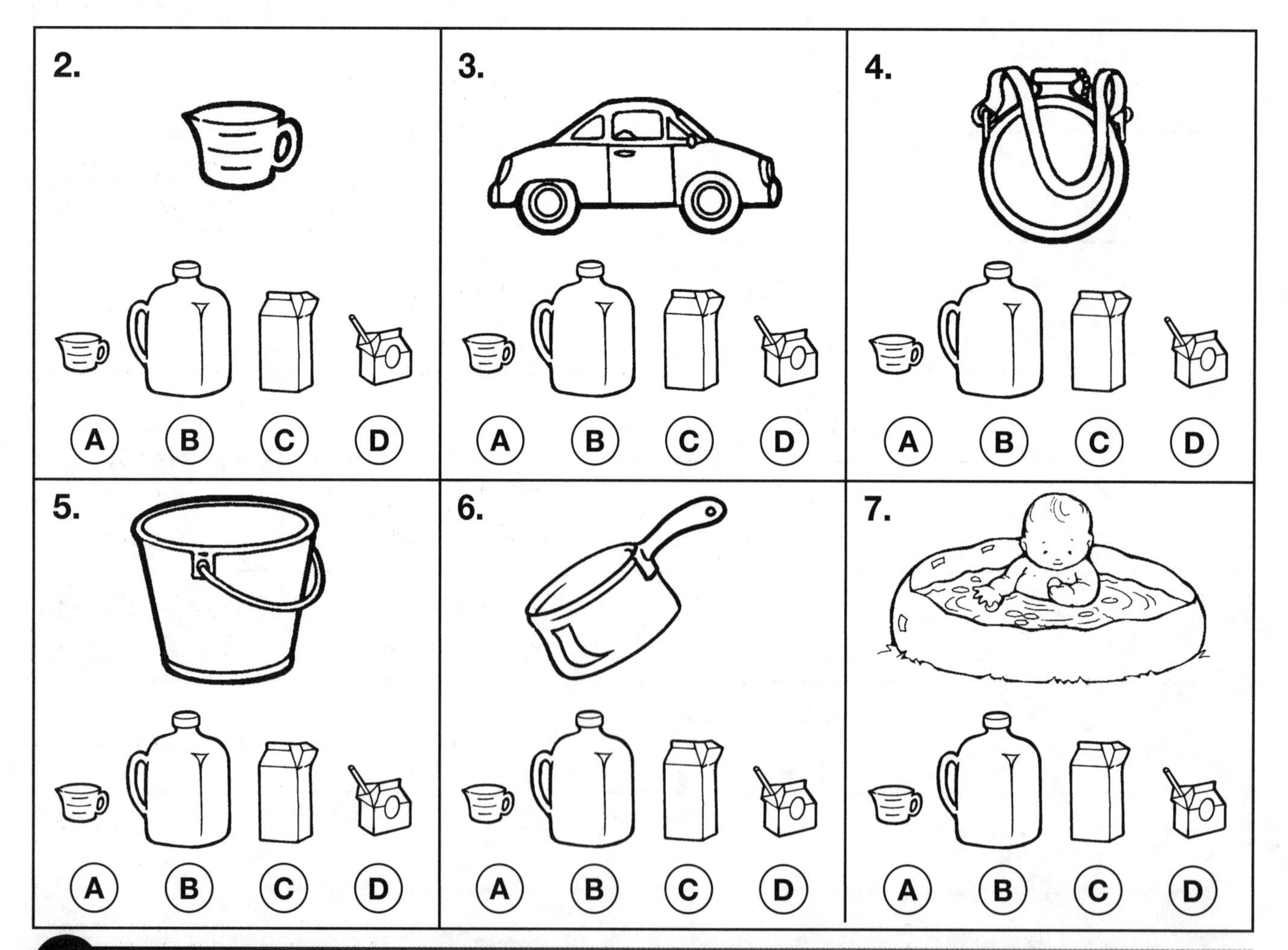

Practice 28

When cooking, small amounts of ingredients can be measured in cups and pints. Complete the table. (Note: 1 cup = ½ pint)

Cups	1	2						8
Pints	1/2	1	1½			3		

Use the symbols >, <, or = to compare the different amounts.

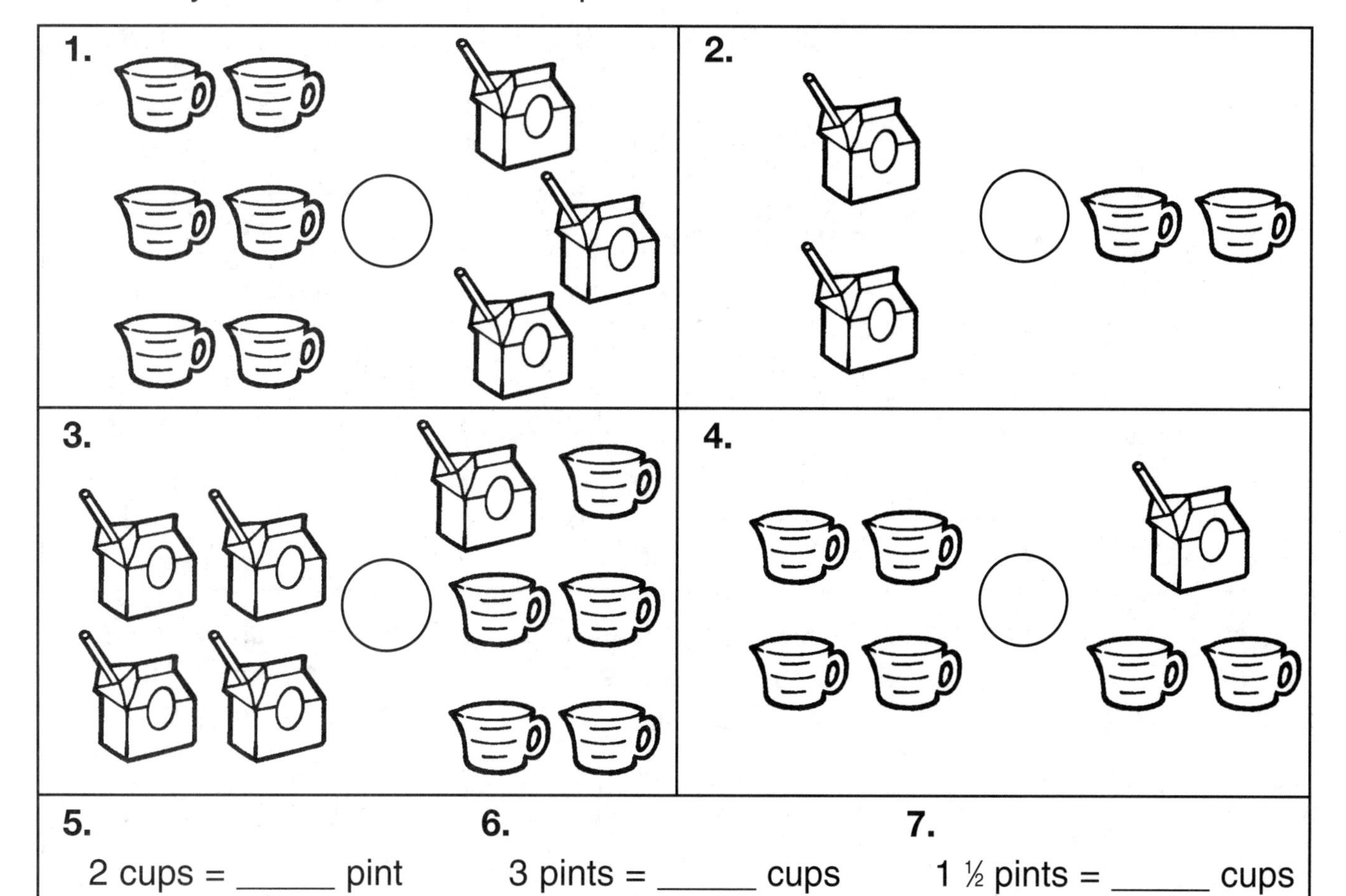

5. 2 cups = ______ pint

6. 3 pints = ______ cups

7. 1 ½ pints = ______ cups

Practice 29

Cups, pints, quarts, and gallons are used to measure liquids and dry ingredients.

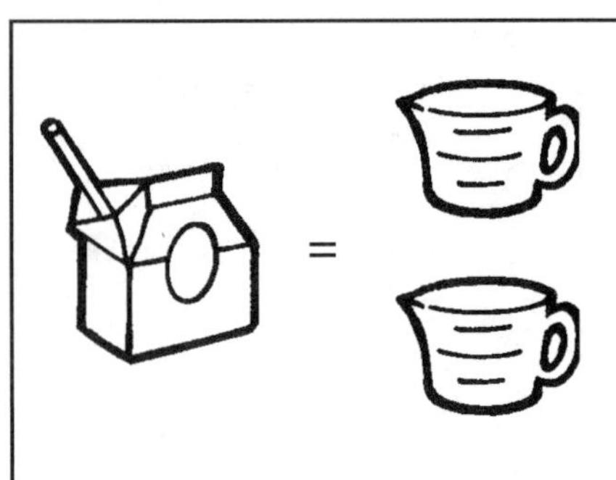		
1 pint = 2 cups	1 quart = 4 cups	1 gallon = 16 cups

Answer each question.

1. About how much water could a large mixing bowl hold? 5 quarts (A) 5 pints (B)	**2.** About how many gallons would fill a sink? 50 gallons (A) 5 gallons (B)	**3.** How much detergent should you use to wash clothes? 1 cup (A) 10 cups (B)
4. How many servings of punch are in one large pitcher? 8 cups (A) 8 quarts (B)	**5.** How much milk could a baby drink at one time? 1 pint (A) 1 gallon (B)	**6.** How much ice cream should be bought for a family of 4? 1 pint (A) 1 quart (B)
7. How much soil should be put in a small planting pot? 2 pints (A) 2 gallons (B)	**8.** How much water would you use to wash a car? 10 gallons (A) 10 gallon (B)	**9.** How much water would you use to brush your teeth? 1 gallon (A) 1 cup (B)

Practice 30            

Cups, pints, quarts, and gallons can be used to measure different items such as cooking ingredients, water, gasoline, or sand. Complete each set of directions with the correct measuring container.

1. When the car is out of gas, carefully fill it with a __________ of gas.

  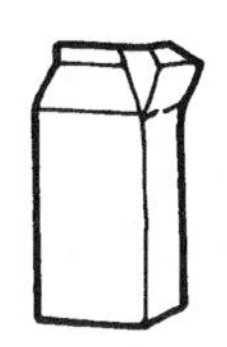 

cup (A) pint (B) quart (C) gallon (D)

2. Instead of making ice tea by the glass, prepare it in a pitcher by the __________.

  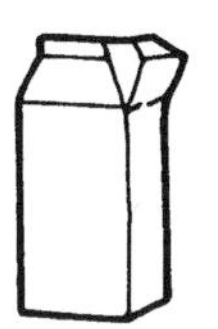 

cup (A) pint (B) quart (C) gallon (D)

3. To make a sundae, scoop out 1/2 quart of ice cream or use a full __________ instead.

cup (A) pint (B) quart (C) gallon (D)

4. Instead of carrying a heavy gallon of cream, put the cream into four __________ cans.

  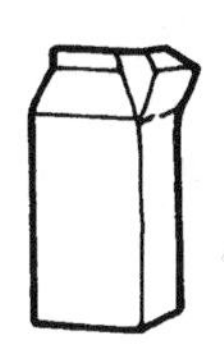 

cup (A) pint (B) quart (C) gallon (D)

5. When eating cookies, carefully dunk the cookies into a __________ of milk.

  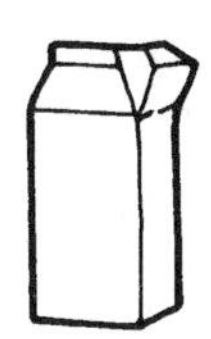 

cup (A) pint (B) quart (C) gallon (D)

6. When making punch, pour one packet of mix into a pitcher that holds a __________ of water.

  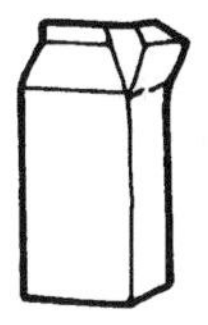 

cup (A) pint (B) quart (C) gallon (D)

Practice 31

It takes about 4 liters to equal 1 gallon.

4 liters = 1 gallon

Follow the directions. Fill in the circle under the correct picture.

1. Holds more than 1 liter.

Ⓐ Ⓑ Ⓒ

2. Holds less than 1 liter.

Ⓐ Ⓑ Ⓒ

3. Holds about 1 liter.

Ⓐ Ⓑ Ⓒ

4. Holds less than 1 liter.

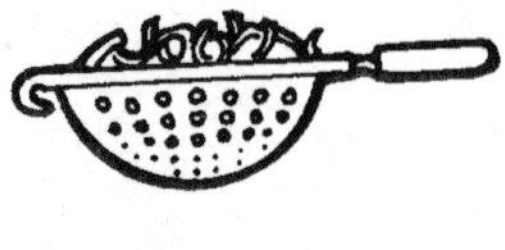

Ⓐ Ⓑ Ⓒ

5. Holds about 1 liter.

Ⓐ Ⓑ Ⓒ

6. Holds more than 1 liter.

Ⓐ Ⓑ Ⓒ

Test Practice 1

1. About how many paper clips long is the snake?

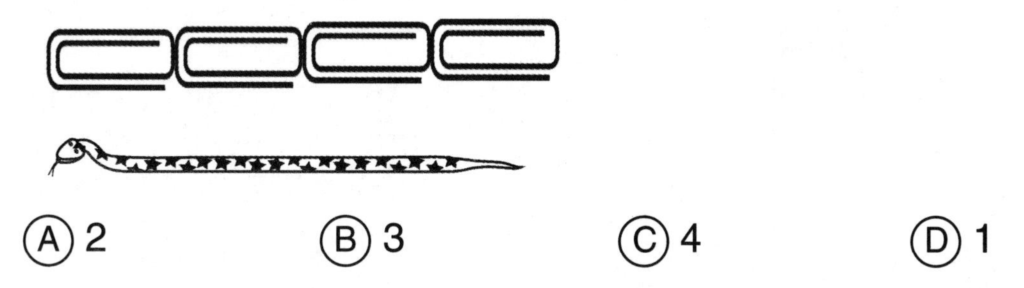

Ⓐ 2 Ⓑ 3 Ⓒ 4 Ⓓ 1

2. Measure the item. About how long is it?

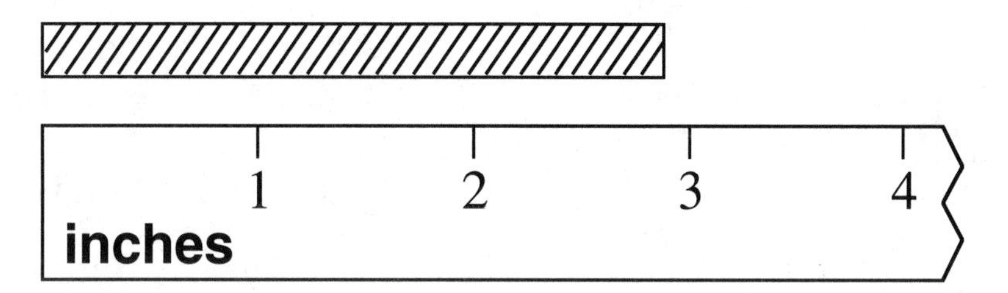

Ⓐ about 3 inches Ⓑ about 2 inches

Ⓒ about 1 inch Ⓓ about 4 inches

3. Which group lists the items from *shortest* to *longest*?

Ⓐ a chalkboard, an eraser, your shoe, your hand

Ⓑ an eraser, your hand, your shoe, a chalkboard

4. About how many paper clips long is the baton?

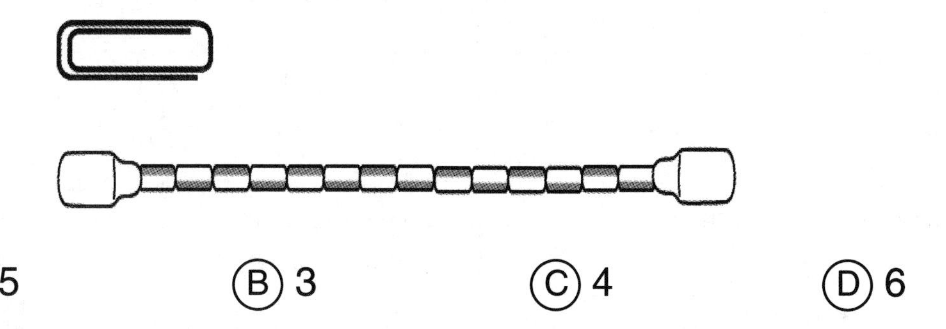

Ⓐ 5 Ⓑ 3 Ⓒ 4 Ⓓ 6

5. Which group lists the items from *longest* to *shortest*?

Ⓐ a pencil, your shoe, a chair, a boat

Ⓑ a boat, a chair, your shoe, a pencil

Test Practice 2

______ = 1 unit

1. About how many units is it from the apartment building to the basketball court?

   1 unit (A)   2 units (B)   3 units (C)

2. About how many units is it from the tree to the bench?

   1 unit (A)   2 units (B)   3 units (C)

3. How many inches long is the snake?

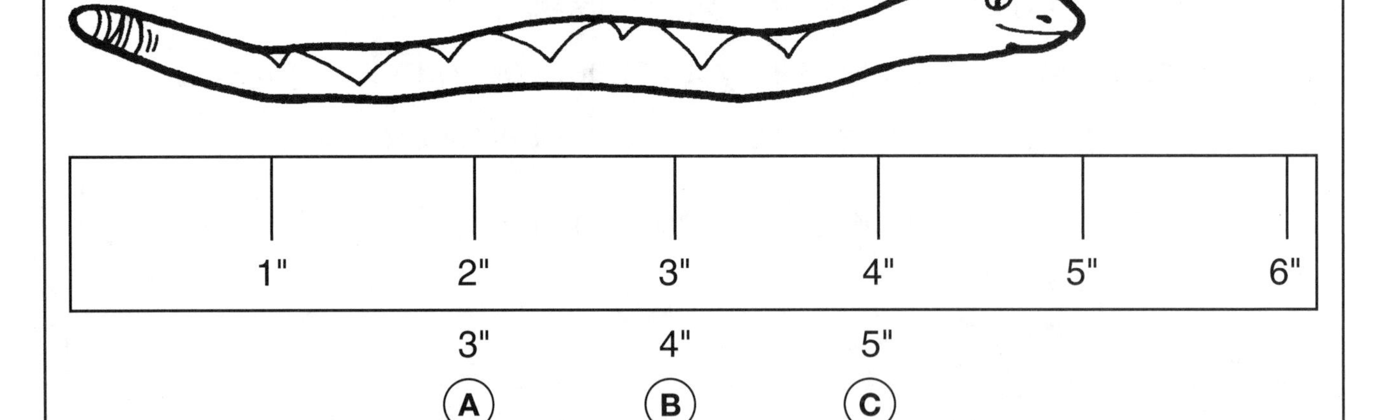

3" (A)   4" (B)   5" (C)

4. About how long is a bee?

   (A) 1 inch

   (B) 1 foot

5. About how tall are some trees?

   (A) 6 inches

   (B) 6 feet

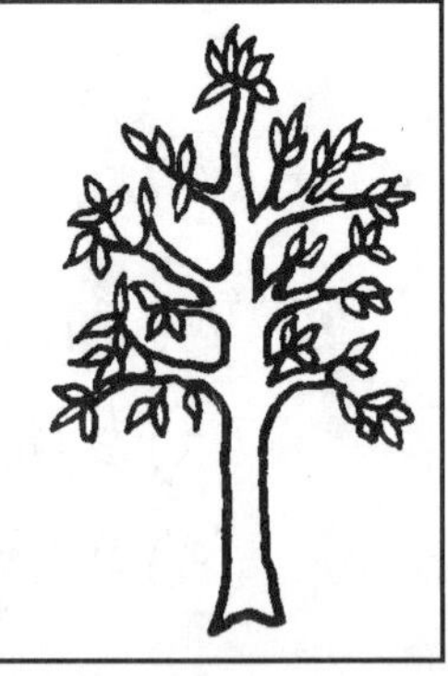

Test Practice 3

1. Measure the wrench to the nearest centimeter.

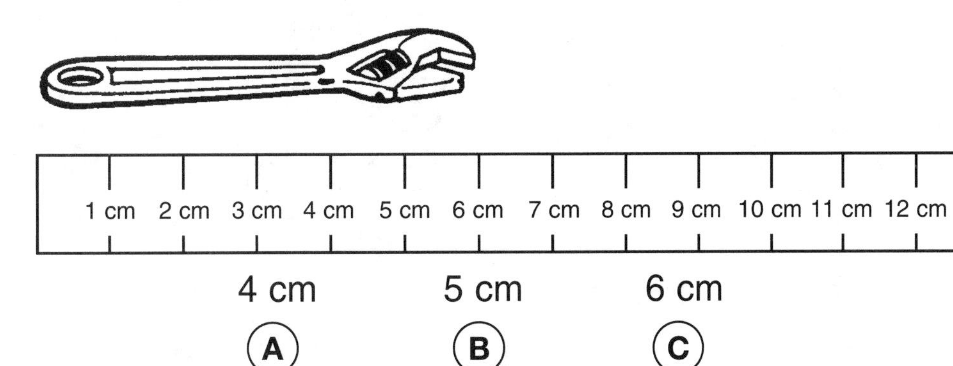

4 cm (A) 5 cm (B) 6 cm (C)

2. Would you measure a mouse in centimeters or meters?

centimeters  meters 

3. Would you measure a giraffe in centimeters or meters?

centimeters (A)  

4. About how long is this snail?

1 cm (A) 2 cm (B) 3 cm (C)

5. About how long is this pack of gum?

1" (A) 2" (B) 3" (C)

6. Which circle shows the length of an object 4" in length.

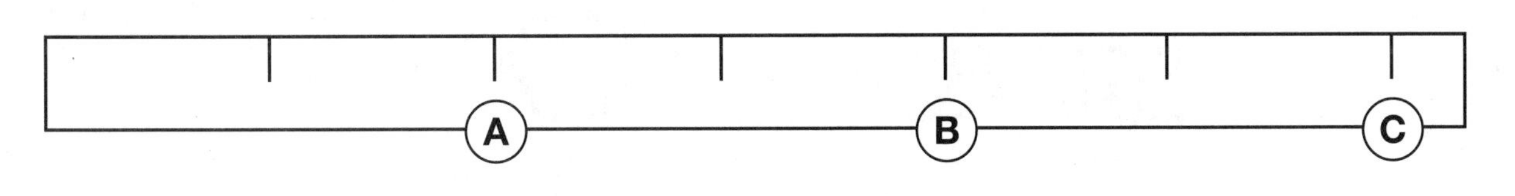

Test Practice 4

1. How much would the container hold?

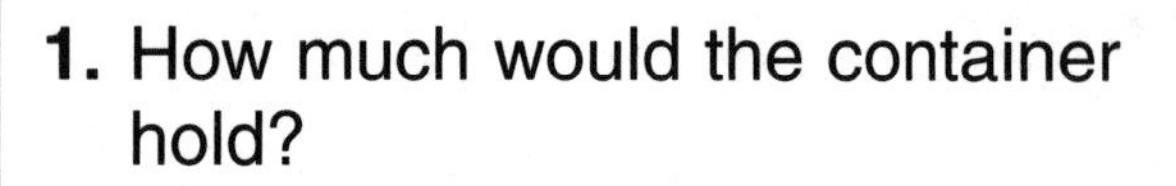

1 pint	1 cup	1 gallon
(A)	(B)	(C) 

2. Which symbol should be used in the circle?

(A) > (B) < (C) = 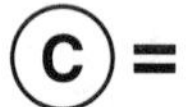

3. Which symbol should be used in the circle?

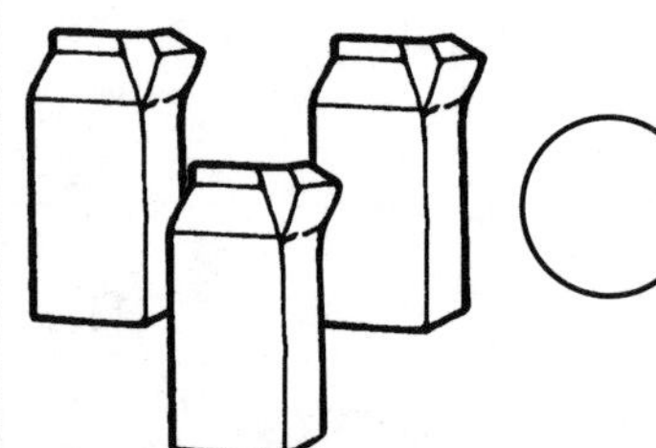
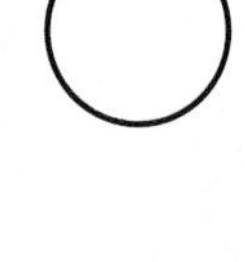

(A) > (B) < (C) =

4. Which symbol should be used in the circle?

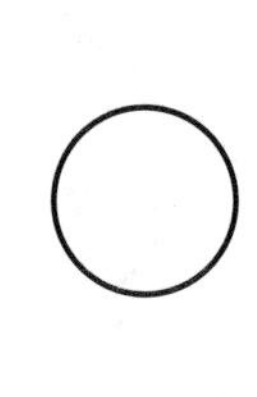

(A) > (B) < (C) = 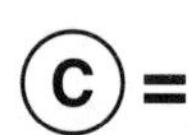

5. Which container would hold the most water?

(A) (B) (C)

6. Which would you use to measure a cup of flour?

(A) (B)

7. What would you use to measure the length of a shoe?

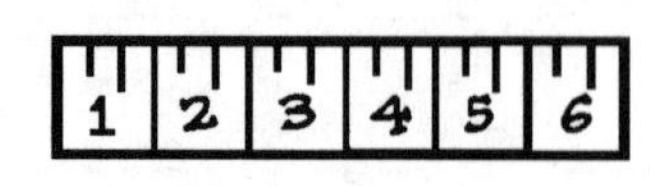

(A) (B)

Inch Rulers and Yard Sticks

Make a 12-inch ruler by cutting out the first two strips and taping the end of the first strip to the beginning of the second strip. To make a yard stick, cut out all the strips and tape them end to end so that the numbers are in order from 1 to 36.

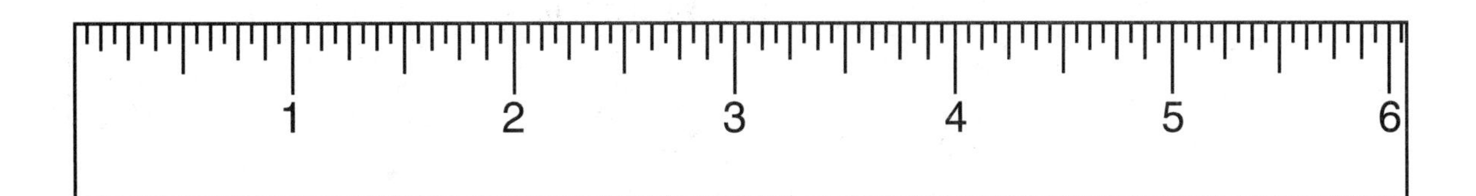

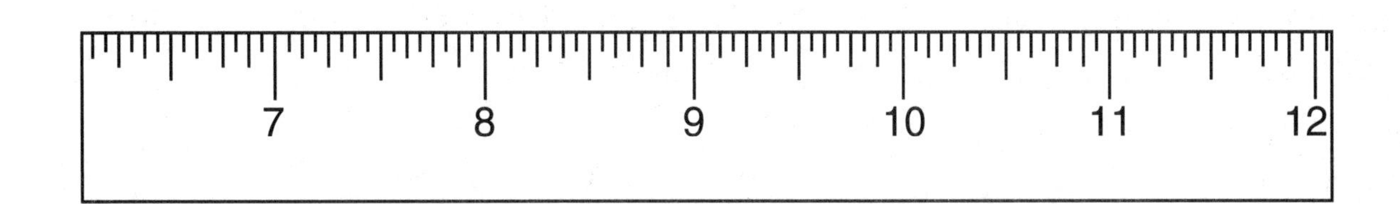

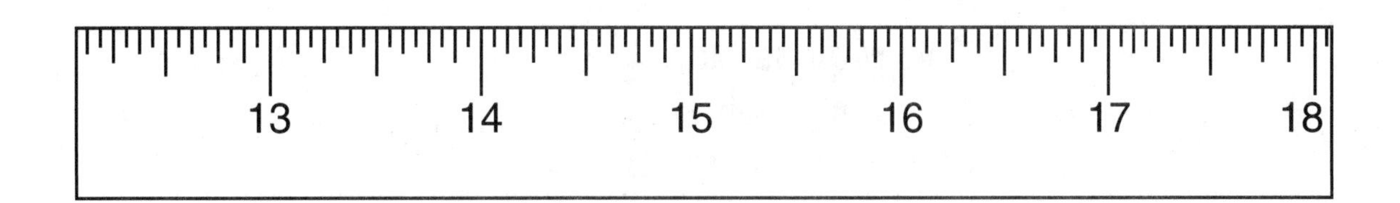

25 26 27 28 29 30

31 32 33 34 35 36

This page is blank
for cutting out
objects on page 39.

Centimeter Rulers

Use the first strip for some of the activities in this book. A meter ruler measures 100 centimeters and is about three inches longer than a yard stick. To make a meter ruler, cut out the strips and tape them end to end so that the numbers are in order from 1 to 100.

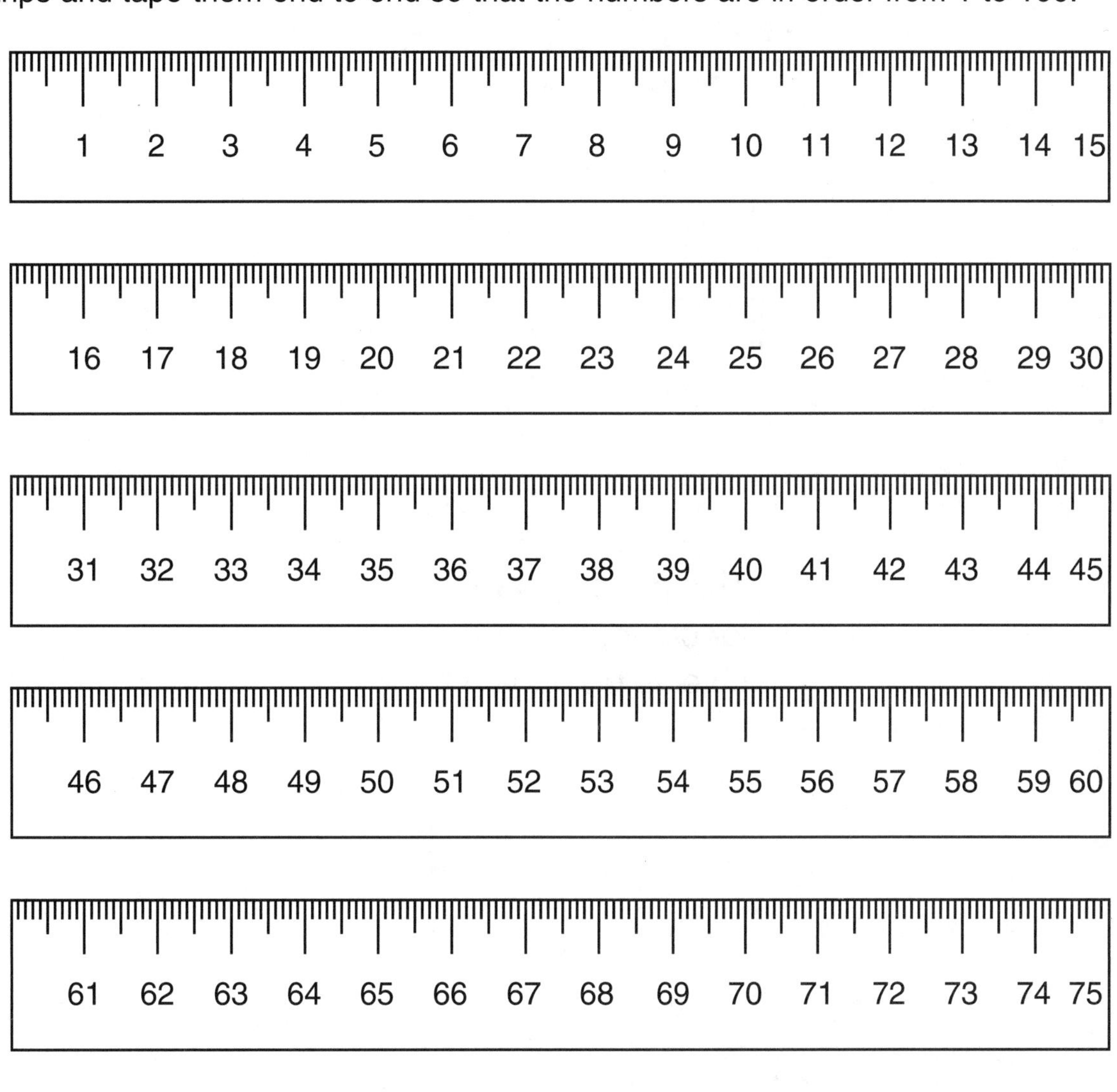

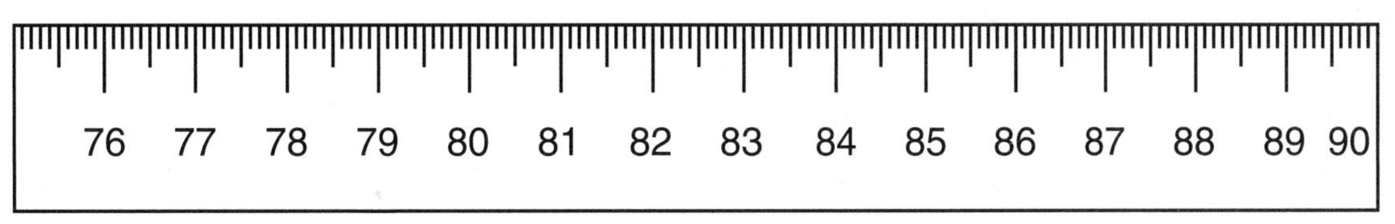

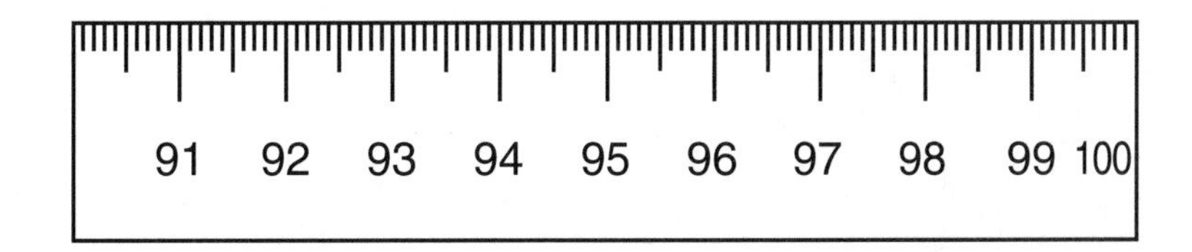

This page is blank
for cutting out
objects on page 41.

Cut Outs

The blocks and inchworms below can be cut out and used for some of the activities in this book or to practice measuring items of your choice.

This page is blank
for cutting out
objects on page 43.

Cut Outs           

The paper clips and buttons below can be cut out and used for some of the activities in this book or to practice measuring items of your choice.

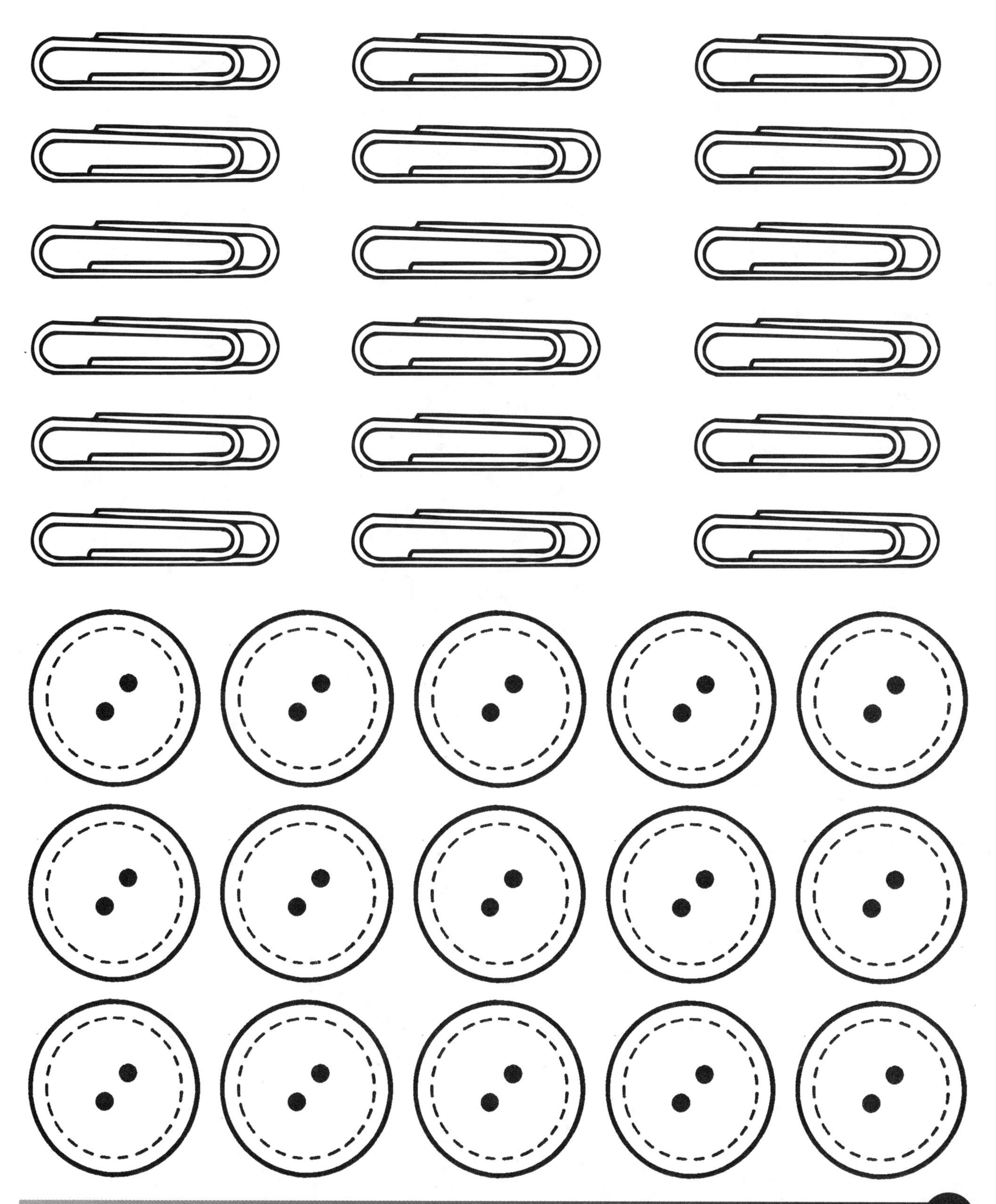

This page is blank
for cutting out
objects on page 45.

Answer Sheet

Test Practice 1	Test Practice 2
1. Ⓐ Ⓑ Ⓒ Ⓓ	1. Ⓐ Ⓑ Ⓒ
2. Ⓐ Ⓑ Ⓒ Ⓓ	2. Ⓐ Ⓑ Ⓒ
3. Ⓐ Ⓑ	3. Ⓐ Ⓑ Ⓒ
4. Ⓐ Ⓑ Ⓒ Ⓓ	4. Ⓐ Ⓑ
5. Ⓐ Ⓑ	5. Ⓐ Ⓑ

Test Practice 3	Test Practice 4
1. Ⓐ Ⓑ Ⓒ	1. Ⓐ Ⓑ Ⓒ
2. Ⓐ Ⓑ	2. Ⓐ Ⓑ Ⓒ
3. Ⓐ Ⓑ	3. Ⓐ Ⓑ Ⓒ
4. Ⓐ Ⓑ Ⓒ	4. Ⓐ Ⓑ Ⓒ
5. Ⓐ Ⓑ	5. Ⓐ Ⓑ Ⓒ
6. Ⓐ Ⓑ	6. Ⓐ Ⓑ
	7. Ⓐ Ⓑ

Answer Key

Page 4

1. C
2. B
3. B
4. A
5. C

Page 5

1. B
2. B
3. clock, sofa, whale
4. giraffe, table, candle

Page 6

1. C
2. D
3. B
4. A

Page 7

1. 2
2. 3
3. 3
4. 2
5. 3
6. 6

Page 8

Answers will vary.

Page 9

1. B
2. A
3. C
4. Answers will vary.
5. Answers will vary.
6. A
7. C
8. B
9. Answers will vary.
10. Answers will vary.

Page 10

1. 2
2. 1
3. 2
4. 1
5. 1
6. 2
7. 2
8. 3
9. 3
10. 1

Page 11

1. A
2. Path 2
3. Path 3
4. 5 units

Page 12

1. Juan
2. Eddie
3. Josh, Tami
4. Kim's
5. Kim

Page 13

1. A
2. 5
3. 3
4. 4
5. 5
6. 2

Page 14

Color bubbles and YES boxes yellow for 1, 2, 3, 9, 11, 13.

1. 6
2. 7
3. cannot, 1

Page 15

1. 11
2. 8
3. 14

Page 16

1. 2
2. A
3. 3
4. 3 inches

Page 17

1. 3, 3
2. 1, 1
3. 4, 4
4. 5, 5
5. 2, 2
6. 6, 6

Page 18

1. 3
2. 1
3. 4
4. 5
5. 2
6. 6
7. 3
8. 4

Page 20

1. 4
2. 3
3. 2
4. 5
5. less
6. less
7. less
8. less
9. more
10. less

Page 21

1. 2
2. 1
3. 5
4. 4
5. 6
6. 3

Page 22

1. 4
2. 6
3. 3
4. 6
5. 4
6. 8
7. 6
8. 1

Page 23

1. 4
2. 1
3. 10
4. 7
5. 11
6. 3
7. 8
8. 5

Page 24

1. D
2. D
3. C
4. C
5. C

Page 26

1. 5 cm (actual)
2. 12 cm (actual)
3. 7 cm (actual)
4. 7 cm (actual)
5. ladybug #2

Page 27

1. shorter
2. shorter
3. taller
4. shorter
5. taller
6. taller
7. shorter
8. shorter
9. taller

Page 28

1. m
2. cm
3. cm
4. cm
5. m
6. cm
7. m
8. m
9. m

Page 30

1. 3, 1, 4, 2
2. A
3. B
4. D
5. B
6. C
7. B

Page 31

Cups	1	2	3	4	5	6	7	8
Pints	1/2	1	1½	2	2½	3	3½	4

1. =
2. >
3. >
4. =
5. 1
6. 6
7. 3

Page 32

1. B
2. B
3. A
4. A
5. A
6. B
7. A
8. A
9. B

Page 33

1. D
2. C
3. B
4. C
5. A
6. C

Page 34

1. C
2. B
3. A
4. C
5. C
6. A

Page 35

1. C
2. A
3. B
4. C
5. B

Page 36

1. B
2. C
3. C
4. A
5. B

Page 37

1. C
2. A
3. B
4. B
5. B
6. B

Page 38

1. B
2. B
3. B
4. A
5. C
6. B
7. A